Absolute Space and Time

Absolute Space and Time

A New Approach to the Speed of Light

by

Francis Pym and Clifford Denton

Twoedged Sword Publications

First published 2005 under the Title *Einstein's Predicament*. Now revised and updated.

ISBN-13   978-1-905447-22-0

ISBN-10   1-905447-22-1

Twoedged Sword Publications
PO Box 266, Waterlooville, PO7 5ZT
www.twoedgedswordpublications.co.uk

Foreword

The subject of Relativity is fundamental to the mental-paradigms that control our thinking in so many areas of life, that any challenge to its basic tenets has very far-reaching consequences. Even a non-scientist such as me can appreciate this much. What is especially important about the work of Francis Pym and Clifford Denton in their book *Absolute Space And Time* is that their proposals are so elementary.

To take but one example from their work; in any experiment that might be set up, we may only measure the average speed of light over its outward and return journeys *combined*, but how often do we acknowledge the crucial nature of this basic fact? When this truth and its consequences are set before us in such an illuminating manner, we begin to realise that hitherto we have failed to think about the key issues with any real clarity.

The two writers stimulate our minds immensely by presenting us with the very first First Principles that we have to deal with in a way which grips our imagination and fires our enthusiasm. Inevitably, we are left wondering quite where it will all end in an age that is wedded to Post-Modernism's anti-absolutist world views, but these little concerns must never be allowed to smother such an important enquiry. To say this book is challenging is seriously to downplay it; it is a coherent, intelligent and lucid invitation to reconsider the entire meaning of life; because real evidence for the genuine existence of absolutes would change *everything*.

I am not qualified to comment on the mathematical and scientific details of the arguments advocated by the authors, but they have certainly produced the most stimulating read I have encountered in a long, long time.

I will not be the only one of their readers left feeling grateful to them for letting in such a breath of fresh air, that the cobwebs clinging to our normal modes of thinking have all been blown away! I will not spoil the reader's enjoyment by letting any cats-out-of-the-bag, but the way the authors handle the equation that has been associated with Einstein for decades, namely $E=mc^2$, is especially intriguing. They show with a simple classical argument that this famous equation is quite independent of Einstein's Theories. Many a reader will be surprised and illuminated by the authors' helpful comments at this point.

I feel deeply honoured to have been asked to write the Foreword to such an exciting and vitally important piece of original work as this. May it be a springboard to further endeavours along these lines, and may it initiate a genuinely open debate on the entire subject. It has the potential to be an opinion-changing, landmark work.

M. W. J. Phelan PhD
Sussex, December 2008

Acknowledgements

Francis Pym

There are many to whom I owe much in preparation for this book. In early days I was given a book on Relativity and was struck by what I saw as a lack of logic in the argument. In thinking of an alternative together with Dr Lindsay Douglas, I was introduced to Dr Adrian Otterwill and others at the Dept of Mathematics, Magdalene College, Oxford.

Dr Clifford Denton my co-author, without whom little progress could have been made then helped in bypassing Relativity Theory. His belief in what we were doing was important, as was our bond and our untiring tenacity in searching together for the truth. He pressed for progress using logical engineering procedures taking one step at a time and using language that made sense in expressing the emerging intuitive ideas.

My thanks also go to the Trustees of the Joshua Trust for their interest, encouragement and support; to Mrs Rosabel Topalian for her role as a prudent scholar of physics in guiding our thoughts and correcting elementary mistakes; to Dr David and Joan Rosevear for their support and for inviting me to lecture at their 7th European Creation Conference where I presented my paper entitled "2-fluid Theory. The Physical ⟨*Summer* Implications of a Background Medium in Space"; to Mrs 2000 Olga Marshall for her hospitality, understanding and wisdom during conferences; to Dr Peter Holmes for encouragement to persist in reaching out into the topic;

and others also for their less favourable but critical and helpful analyses.

Our son John with his aptitude with equations and for his professional civil engineering skills, his sharing and contributing his own ideas for experiments and so on, and our daughters Rebekah and Victoria were all curiously interested and made helpful suggestions. Many others read the material and saw that what we were doing made sense. Our thanks also go to our interested and patient publisher Paul Rose of Twoedged Sword Publications, introduced to us by Mike Phelan.

Last of all but by no means least I acknowledge with precious thanks, my wife Marigold; for her loving support over a long period when she could not have understood much of what we were doing; for helping with our English, proof reading and so on, and most of all JCOL.

Francis Pym
January 2009

Clifford Denton

I am grateful to have studied mathematics and physics in an age where absolute frames of reference were assumed, being then able to study the basic principles of Relativity Theory from this perspective. It gave me a caution when approaching theories which promised much but seemed to expect too much of a mindset change to accept them wholesale.

My first acknowledgements, therefore, go to my teachers who cultivated an interest in the laws of Physics and the tools of Mathematics, so setting a firm foundation in my academic life. For those born into an age of relative thinking, I would propose that they do not know what they are missing, compared with those of us with a secure framework cultivated from absolute principles in law and morals. It is for this reason that I have desired to understand the workings of the universe and, since meeting my colleague Francis Pym, of investigating what we perceived as the errors of Relativity Theory.

In our partnership, I have gained immense respect for the tenacity of Francis in not giving up and so keeping this project alive. Put a mathematician together with an architect and there is both the potential for what we have now achieved or the potential for sleepless nights due to the different means of communication. We have wrestled with how to express ideas, born mainly from intuitive concepts, into mathematical language, albeit that the mathematical tools used were not complex. I greatly acknowledge the contribution of Francis to keep this project alive and to see it through to completion.

The concepts we have wrestled with did not require complex mathematics but they—the concepts themselves—are not as easy to grasp as we would have hoped. We needed to enter the mindset of relativity in order to seek to correct it, and this is something that our readers will have to face too. I have also greatly valued the many interactions that have taken place in the home of Francis and Marigold, the hospitality, the discussions on other topics of Christian apologetics, and the environment of faith and prayer that we find there.

Finally, for my part, there is the acknowledgement of encouragement from others of our family and friends who have shown interest in the end results of our investigations, who never quite understood nor believed in Relativity Theory, but couldn't say why. It is to these that we now offer this book.

Clifford Denton
January 2009

Note on the Title of the Book

The central ideas of this book were first published under the title *Einstein's Predicament*. We chose that title to emphasise the predicament that Einstein faced in investigating the passage of light through the universe, out of which came the Special Theory of Relativity. Our book challenged Einstein's basic assumptions and made a case for a return to absolute measurements in space and time.

While the central ideas have been carried over into this book, there is also sufficient change for us to think a new title to be appropriate. We have also improved some of the arguments, especially those relating to length contraction and reciprocity. A substantial section relating to a two-fluid model of the universe has been transferred to another book entitled *Aether Physics*. We have also added a simple derivation of $E=mc^2$ based on the principle of a static background medium in the universe.

With all these changes, we felt that this was more a new version of the book than a new edition. For this reason, we have given the book a new title, which relates to our objective of highlighting the probable errors in the Special Theory of Relativity. We felt these were best addressed by a return to the foundations of Physics that were built on absolute measurements in the universe.

Contents

1

Concepts

Assumptions and their Consequences

A hundred years ago Einstein published a foundational paper on the subject of light that has largely been accepted as a remarkable working document in the scientific world until now, even though it has been found incompatible with other theories. Most people do not have the scientific background either to understand or to challenge Einstein's papers and have had to put their trust in scientists. Yet, with some simple maths we plan in this book to demonstrate that Einstein may have made some serious errors. These errors are not so much within the development of his papers, as in the foundational assumptions that were made. Therefore, we are seeking to look at the logical implications of bypassing Einstein's theory of relativity which, if we are honest, no one completely understands.

Many people are not trained as scientists so might assume that eminent scientists discover truth. However, the truth of scientific discovery is only as valid as the truth of the founding principles and assumptions. Science such as physics is concerned with observation, experimentation, measurements, framing of theories and predictions, but these things are not examined in isolation. Every scientific conclusion depends on basic assumptions in just the same way that each branch of philosophy depends on foundational assumptions. Scientific and philosophical logic is of the kind; 'if A is true and B is true then we can conclude that C is true'. Science has experimental or mathematical logic to support it so that it can deal with the real world rather than just ideas, as is done in philosophy, though the two disciplines are converging in our day. For Einstein, the assumptions of Relativity Theory were based on the speed of light. [1]

An informed layman with school level maths should be able to follow our reasoning in the rest of the book. However, we would point out that though the mathematics is straightforward the concepts could be quite difficult, so that the book will need to be read and thought through very carefully. This explanatory chapter therefore sets out the main ideas in a more popular manner without the use of mathematics, as a help to the less confident readers, though the remaining chapters may broadly be understood by skipping the maths.

History

Prior to Einstein there had been the idea that the universe was filled with a motionless substance or background medium that was called the *Aether*. Its characteristics in the universe were found hard to assess and its presence hard to detect. However, in this book we introduce a new term for this background medium. We will use the term *Absolute Frame of Reference* (or AFR) rather than *Aether* for technical reasons, but for simplicity in this chapter, we will retain the term *Aether* for the background medium.

When serious consideration was being given to the existence of the Aether it was thought that light travelled through it in waves. However, detection of these waves was not easy. With sound waves we can detect them as they travel through the air, but it has not been possible to either detect the Aether or monitor the travel of light waves through it in the same way. This is largely because light travels faster than anything else and this makes experiment difficult.

If the Aether exists, then it is the substance *within* and *between* the physical elements of the universe that interests us, but we do not have instruments to measure it or vessels to contain it. If it exists, atoms of the universe will interact with the Aether as they move through it. Since these are the smallest constituents of matter, we cannot have instruments small enough to make observations or measurements at that level. Thus, we have to make assumptions

The problem that scientists such as Einstein faced was that there seemed to be a very strange result when attempts were made to measure not only the speed of the earth through the Aether but also the speed of light itself. This was not like measuring the speed of sound.

Suppose a vehicle transmits sound waves as it moves. To an observer travelling with the vehicle, the sound transmitted in front will be measured as being slower than when measured by someone who is stationary on the ground. For example, a jet plane can fly through the air at such a rate that it can sometimes exceed the speed of its own sound. Usually a person on the ground will hear the sound of a plane as it passes by, and for a fast plane, its sound appears to come from behind.

For a straightforward situation in a moving vehicle, a person measuring his own sound subtracts the velocity of the vehicle from that of its sound to obtain its apparent velocity.

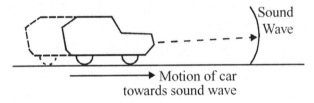

Sound Wave

→ Motion of car
towards sound wave

This did not seem the same for light waves. Whether a body was moving or not, it seemed from the experiments that had been carried out that no subtraction was necessary. Whether moving or not, everyone seemed to get the same answer for the speed of light. This result naturally cast doubt on the existence of the Aether as well as establishing a philosophical problem for scientists. When Einstein began to think about this problem, he proposed that it was possible to avoid it and set out to get round it, in a sense by ignoring it.[2]

To him it was not necessary to refer to a universe that was at absolute rest. If the Aether could be detected it was thought that it would define the state of absolute rest. Since the Aether could not be detected, Einstein *assumed* that by using a new system of relative measurements he could ignore it, so he abandoned the absolute measurements that would have been related to the Aether. Secondly, since the speed of light always *seemed* to be measured as constant whatever the motion of an object through the universe, Einstein established this as the other foundational *assumption* of his Relativity Theory.

While Einstein was able to make great and convincing strides forward, the consequence was that physical quantities then needed redefinition. For example, time became a relative concept, with time changing as a body moved, no longer relying on absolute measurements. Scientists are at liberty to make new definitions that are consistent with the theories, but they also have to bear the consequences. One major consequence of Einstein's theory is the loss of absolute measurements. That said, there is not only the spread of relativity theory into other areas of philosophy but also into the conscience of mankind and the whole area of relative morals.[3] Moreover, there can also be errors in having wrong assumptions that are then carried forward into other major areas of science *(see chapter 8, Addendum)*.

Passage of Light Through Space

We propose that one major error lies in the assumption that light travels at the same constant speed to every observer. As we have said, measurement of the speed of light is not as straightforward as the measurement of the speed of sound. This is because of its immense rapidity and because nothing is known to travel at a greater speed. This means that when we send out a light ray it is impossible physically to keep up with it to see how fast it travels.

One might be tempted to think that the way to overcome this would be to set up an experiment. Suppose a light signal were sent out from a source at point A to be received at point B at a known distance. One could then determine the time of travel in the same way that one would calculate the speed of sound in air leading to a calculation of the speed of light.

However, herein lies a fallacy. One needs to be able to synchronise clocks at A and B first. How can one do this? Either one puts the clocks together and synchronises them before moving them apart, or one sends a signal from A to B so that the clock at B is set to the time at A. We cannot be sure that either of these methods works.

In the first case, when we move the clocks apart, we are not sure if the movement of the clock from A to B changes its time. In fact, as we explain below, we do believe that this can be so. In the second case, the signal that we would send from A to B to synchronise the clocks would have to be transmitted by an electromagnetic signal at a known speed. This would be at the speed of light — the very thing we are trying to measure.

Thus, every scientific experiment to measure the speed of light has relied on a different approach from this. In these experiments, light is sent from a transmitter at A to a reflector at B and back, while the light is timed over the double journey. The point is that all attempts to measure the speed of light that have produced useable results, give an *average speed for this double journey* rather than an actual measurement of the speed of light in one direction.

Let us take an analogy from the measurement of the speed of sound in air. Suppose a vehicle is travelling towards a distant wall at a certain speed and sends out a sound wave carried in still air.

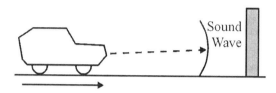

The sound travels in the air and the vehicle moves a certain distance before the sound reaches the wall and the echo returns. The speed of sound on the outward journey as measured from the vehicle is the actual speed of sound in still air *minus* the speed of the vehicle.

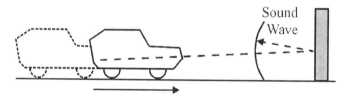

On the return journey, since the vehicle is moving towards the echo, the apparent speed of sound will be the speed of sound in still air *plus* the speed of the vehicle.

If both the vehicle *and* the reflecting wall moved at the same speed, maintaining the distance between them or if there were a head wind, the effect would be amplified. In this case, we could have calculated the average speed of sound by dividing the 2-way distance by the total time. Again, we would not have taken account of the fact that, measured in the vehicle, the sound travelled at a different speed in each of the two directions, i.e. to the wall and back.

Now, this is a point that has never been refuted even in Einstein's Theory of Relativity. If light travels like sound in the medium (the Aether) as we propose, then light too travels at a different speed in each of the two directions as measured by moving instruments. If the light source and reflector, A and B, are travelling together through the medium there will be different speeds to and from the reflector that are 'averaged out' when a calculation is made of the 2-way journey. The problem here is that in the case of the measurement of the speed of light we cannot make the two separate measurements to and from the reflector and so detect if there is an error in our assumption.

However, we do show a simple calculation later in the following chapters that gives a remarkable result in the case of the measurement of the speed of light. This is for a 2-way journey from source A to reflector B and back. If A and B are moving at any speed through the medium then we always get the same 'average' speed for light for the double journey even though the speeds of light to and from the reflector are themselves different. Thus Einstein's assumption *seems* correct, but it is only correct for an averaged 2-way passage of light. Moreover, this result cannot be used to infer (as it is in modern science) that light always travels at the same speed whether the instrument is moving through the universe or not.

Such suppositions have important consequences when we consider light from distant stars moving with respect to us, since the stars themselves are moving as well. When we come to consider the age and development of the universe, assumptions about the speed of light have major implications. One such of course is the contribution this has to the theory that the universe began with a 'Big Bang'.

Einstein's assumption about the speed of light allowed him to proceed with the Theory of Relativity which then became a self-contained mathematical system. To proceed with his theory other issues needed reinterpretation, for instance the measurement of length and time in a moving body. In practice, the equations of Special Relativity relate to bodies moving at very high constant speeds requiring that time slows and lengths shorten as a body moves. They start with the assumption that light always travels in any direction at the same speed according to any observer. Conclusions about change of time and length followed afterwards.

Concern

We argue the case differently. We are very concerned about Einstein's relative measurements in his view of the universe and at the loss of absolute measurements. To Einstein, instruments in each body (space ships, planets, the earth, the sun or stars, for example) make all measurements relative to the body in which they are located, with no reference to any absolute measurements. Each body in the universe, to Einstein, is like its own world unto itself with no need to refer to a set of absolutes that must be defined in some way. His theory came about because of the inability of scientists to detect the Aether that would define the position of an absolute rest in the universe from which all absolute measurements could be made. However, in abandoning the concept of absolute rest, one's concept of the universe changes too.

We propose a return to the concept of absolute rest and absolute reference for time and length. This is even though the Aether has not and probably cannot be detected. The issue here is truth. If we cannot detect the Aether, this is no reason to ignore the possibility of its existence. In the end such theories as Relativity Theory introduced confusing paradoxes and mystical ideas about the universe compared with logical alternatives based on the return to absolutes.

There is therefore a reasonable case for leaving the Theory of Relativity and building afresh on absolutes. Of course, we still need to deal with such issues as slowing of time and shortening of lengths. We show later that shortening of lengths can be explained in a different way. We propose that the speed of forces between molecules, that hold matter together, varies in relation to the speed that the matter travels through the aether. It is this variation of the speed of forces that changes lengths. In regard to time, we would simply state that it is not *time* that changes with motion. It is the rate at which *clocks* record the passage of time that changes due to the motion of the matter itself.

Simultaneity

The final important point to raise in this chapter is the notion of something called 'Simultaneity'. To proceed with his theories, and knowing the impossibility of synchronising clocks at two different points, A and B, since one has to send a signal from A to B to do this, Einstein cleverly bypassed the problem. He used the word 'Synchronicity' instead of 'Simultaneity' and *defined* this term for the purposes of his new theory. He could not solve the problem of Simultaneity so defined himself out of the problem! Even if two events in relation to absolute measurements in the universe did *not* occur at the same time in relation to absolute measurements, Einstein could still *define* them as 'synchronous' according to his theory.

Remarkably, in spite of redefining certain other issues such as time and simultaneity and failing to disprove the existence of a background medium in the universe, Einstein's theory has held ground for a hundred years. We suggest therefore, that it is now timely to challenge this approach through the return to absolutes, and hold out the possibility of returning to this more easily understood basis of measurement in the universe.

2

Proposal

In this book, we reconsider the passage of light with respect to what we will call an Absolute Frame of Reference (AFR) at absolute rest in the universe. Was Einstein right in his supposition that some sort of physical matrix or AFR in space can be ignored without consequences? He redefined certain issues including time and simultaneity in order to proceed with his Theory of Special Relativity (SR). This approach, though attractive, is not consistent with theories based on absolute measurements, and bears consequences that cannot be neglected. By re-examining Einstein's second postulate in SR, we propose that an AFR is plausible, although its location may be uncertain. In reopening this route of inquiry, we suggest a return to the basis of an AFR model and to the investigation of the original meaning of time within a universe of absolutes.

To do this we reconsider some of the foundational issues addressed in the early papers on SR. We then open the way for an alternative model, showing that the prevalence of an AFR is consistent with experimental results that caused its neglect in the derivation of SR. The authors do not dispute that the results of SR have been outstanding as far as advance· in science is concerned. We propose that the valid results of SR can be confirmed in a new theory of absolutes, but that invalid results must be corrected.

Relativity

Towards the end of the 19th century, scientists faced a dilemma. All attempts had failed to detect an aether in space in which light was thought to travel. Until this dilemma could be resolved or until an alternative route of scientific enquiry could open up, it seemed that progress in understanding the motion of physical bodies would be withheld. In 1905, Einstein's elegant paper, *On the Electrodynamics of Moving Bodies*, broke open a new avenue of thinking and the Theory of Relativity was born.[1] From such a simple starting point, immense strides took place in physics. However, after a hundred years we have come to a new set of dilemmas. While relativity is the standard model for understanding the universe today, Relativity and Quantum Theories still seem incompatible. Rather than going back to re-examine the departure point of a century ago, relativistic concepts are still being included in new theories, such as various String theories.

By its very nature, relativity is not only counter-intuitive but also illogical and foregoes the need for certain absolutes in the universe *(see chapter 4)*. This has considerable implications for the mindset of our day, impinging on philosophy and moral law as well as physics. It also has a bearing on how mankind views the origin of the universe, contributing directly to the idea of a 'Big Bang' versus that which was not by natural means. If relativity is to be retained, a theory such as String Theory has to be imposed, which incorporates concepts of eleven dimensions of space, parallel universes and so on.

When surveying some of the end products of relativity and looking at the propositions of new theories such as these, one might wonder if it might not be better to make a fresh study of the concept of an AFR. In many eyes, returning to the point of departure of 1905 would seem to be a step back. However, let us see if such a step is indeed a step back.

3

Point of Departure

Since 1887, the best known and very accurate experiment by Michelson and Morley (MM) to detect the speed of the earth as it moves through an 'aether' has been repeated many times.[4] A light beam from a source is divided into two rays propagating at right angles to each other, making them travel set distances to mirrors where each is reflected back. Interference patterns produced by the recombined light rays are expected to indicate a time difference that would detect the speed of the earth through this aether. This experiment constantly failed to detect its existence.

Material light aethers in space have never lacked critics. One of the dilemmas was the failure of the MM experiments. How could a substance exist that seemed undetectable or unaffected? Yet, all they had to do was to punch a brick back and forth in the hand to experience the inertia that stems from such matter. The authors propose such a material at rest in what we have called the AFR, a concept in which the material of the aether can move. Thus, if we return to the point of departure, we will find that the hypothesis of an AFR is remarkably helpful.[5]

Nevertheless, the genius of Einstein's new theory lay in a point of departure from this dilemma that seemingly did not need to take account of such an aether.

It is crucial to recall here that *the foundations of SR neither accepted nor denied the existence of an aether.* Prior to Einstein, advance in physics was held up by lack of positive evidence for its existence. Einstein thought he could overcome this by devising a theory that did not depend on it. He therefore set up two assumptions for his theory.[6] The first raised to the level of postulate, that:

> 1) *"not only the phenomena of mechanics but also those of electrodynamics have no properties that correspond to the concept of absolute rest."*

This postulate carried the *assumption* that nothing would be lost in ignoring the light-bearing aether, *while not denying its existence.* The second postulate was,

> 2) *"that light always propagates in empty space with a definite velocity c that is independent of the state of motion of the emitting body."* [7]

Einstein saw that if the speed of light seemed always to be constant he could use the Lorentz Transformations to confirm it.

We repeat this important point, that *in neither of his two postulates did Einstein state that an aether did not exist.* He only accepted that *it could not be detected.* So he moved forward with his theory because *its existence* did not seem to be needed as a matter for study. We quote,

> *"The introduction of a 'light aether' will prove to be superfluous, inasmuch as the view to be developed here will not require a 'space at absolute rest' endowed with special properties."* [8]

Since then, those who have believed in SR may have inadvertently overlooked this point and so ignored the inconsistencies we plan to show in the logic of SR.

Specifically we can state that the application of the second postulate is not consistent with an aether or an AFR. A theory based on the efficacy of an AFR leads to the recovery of the concept of a state of absolute rest in the universe and the consequences of light travelling through a medium at absolute rest.[9] From this we can then build up a view of the physics of the universe based on the use of the medium *whether we can detect it or not*. Furthermore, if any extrapolations of relativity theory unwittingly assume its non-existence, when it does exist, then errors can creep in. Indeed, if one of Einstein's postulates is wrong then there will undoubtedly also be false conclusions from it. By implication, of course, we are questioning the validity of the Lorentz Transformations.[10]

Specific difficulties that we would attribute to SR and therefore to General Relativity would include singularities (matter having mass but no size), aspects of gravitation, and the concept of time. We would also include some issues relating to the age and size of the universe derived from an interpretation of the red shift and brightness of light from distant galaxies.

Our proposition is that Relativity Theory in addition to the mysteries that followed in its train should be replaced.

The alternative would be to recover the hypothesis that when a body moves through space, and when it transmits light, this light is carried with respect to an AFR independent of that body. This is rather like sound that travels through the air separated from its source. All absolute motion is then defined from a rest position in the AFR.

Throughout this book, we consider phenomena relating to the passage of radiation from a classical perspective based on the hypothesis that an AFR itself defines the light. It has also been necessary that reference to SR has been made because of the widespread acceptance of the theory. The purpose in doing so is to draw contrasts between SR and the results of our theory of light. However, it must be said that our line of investigation would have still been the same if SR had not been invented, but of course more difficult. Our primary purpose, therefore, is to demonstrate the plausibility of our theory of light. We nevertheless show that although Einstein created problems through SR, certain results can still be upheld while some of his basic postulates are not supported by our proposal.

4

Einstein's Predicament

We will show in this section that Einstein's definition of 'Simultaneity' will not be consistent with our theory of light. This is because Einstein's supposition is that light travels at constant speed relative to any observer. He applied this principle to solve the problem of the time that light took to travel between two clocks and assumed the time for the outward and return journeys is the same. If this is not the case, as we propose, then his definition of Simultaneity is incomplete. Indeed, this leads to the view that he *redefined time rather than keeping to the conventional concepts of time measurements*, contributing eventually to the idea of contortion of space and time.

In his 1905 paper, Einstein proposed a definition of Simultaneity with a view to establishing the conjecture that not only the phenomena of mechanics but also that of electrodynamics have no properties that correspond to the concept of absolute rest. However, within his definition for Simultaneity lies a predicament, as we shall see.

Now, regarding Simultaneity, Einstein writes: [11]

> *"If there is a clock at point A in space, then an observer located at A can evaluate the time of events in the immediate vicinity of A by finding the position of the hands of the clocks that are simultaneous with these events. If there is another clock at point B that in all respects resembles the clock at point A, then the time of events in the immediate vicinity of B can be evaluated by an observer at point B. But it is not possible to*

compare the time of an event at A with one at point B without further stipulation. So far we have only evaluated an 'A time' and a 'B time', but not a common time for A and B. The latter can now be determined by establishing by definition that the 'time' for light to travel from A to B is equal to the time it takes to travel from B to A. For, suppose a ray of light that leaves from A for B at 'A time' t_A, is reflected from B towards A at 'B time' t_B, and arrives back at A at 'A time' t'_A The two clocks are synchronous by definition if

$$t_B - t_A = t'_A - t_B \text{ "}$$

This is the *definition* of synchronicity that then became the basis in SR for time measurements in all 'frames of reference'. We would propose that Einstein began to *redefine* time and space at this point from relative quantities to an invariant, or space-time interval, τ, and made it subject to the concepts of SR.[12] If we are to re-establish our confidence in an absolute frame of reference (the aether or AFR determining the state of absolute rest), time once more becomes an absolute quantity. This has immense consequences for recovery from erroneous, sometimes strange, hypotheses of SR.

Einstein further stipulates in his definition of Simultaneity[13] that

"Based on experience, we further stipulate that the quantity

$$\frac{2\overline{AB}}{t'_A - t_A} = c$$

be a universal constant (the velocity of light in empty space)."[14]

Einstein assumed that the overall speed of light, as measured by a 2-way passage of signals from A to B, gives the value of the speed of light irrespective of any motion of the observer. *His proposal implies that the time for the outward and return journeys of a light signal between A and B are the same.* His conclusion is therefore theoretical and given without means of proof. This is because practical knowledge of the value of t_B is denied and ambiguity remains. Consequently, if a light-carrying AFR is fundamental, and if the motion of A and B through it does influence the speed of light from A to B and B to A, then we propose that *Einstein chose a definition that does not hold.*

Specifically, the equation

$$\frac{2\overline{AB}}{t'_A - t_A} = c$$

does *not* take into account that in the AFR the time of the passage of light from A to B *can be different* from the passage of light travelling from B to A.

At first sight, the ideas that we present in the following pages may appear to be similar to SR, but they are in fact crucially different. We propose an alternative solution to the problem of the passage of light and synchronicity by examining the effect of motion on the measurement of time and length which, as we shall see, bypasses Relativity Theory.

5

The Effect of Motion on the Measurement of Time and Length

Proposition

In this section, we will consider clock rate change (rather than time dilation) and length decrease in SR.[14] These phenomena are familiar to SR. In SR they are a product of the Lorentz Transformations. However, we propose that these are physical realities resulting simply from the reaction of a body moving in an AFR. While lengths *actually* decrease and clocks *actually* slow, it is important to realise the unquestionable difference from SR. Einstein *redefined time*. We propose that *measuring instruments of time* simply change.

Up to this point of departure, both Coulomb and Maxwell had begun serious studies of non-instantaneous transfer of radiation involving a commodity or dielectric. More recently, Quantum Theory has proposed that molecular systems are held in place by virtual quantum photons that bond over a period at the speed of light. We can picture photons travelling over molecular distances where time of travel is an important criterion. When a body moves through an AFR *the relative velocity of light changes as a result*. This causes clock rate change and length decrease—both real physical effects—that we now illustrate using this change in the speed of light.

Passage of light in the AFR

We can illustrate a line of thought through a simple example on the assumption that ordinary matter depends on the properties of chemical bonding explained in terms of contributions from constituent molecules.

Suppose A, B, and C in the diagram below are three hypothetical centres of bonding held in place in a solid body by these virtual quantum photons. Consider absolute distance AC first. AC is at right angles to the direction of motion of the body that is moving through the AFR towards positions A_1 and C_1 at speed **v**. Due to this motion, the resultant speed of bonding photons apparent to the AFR is calculated from a right-angled triangle of velocities, AA_1C_1.

To instruments at rest in the AFR, the hypotenuse AC_1 of the triangle of velocities AC_1A_1 in the diagram represents the speed for the 2-way passage of photons at **c**. To those moving with centres A and C, it will be calculated over the path A_1C_1 giving a resultant of their speed of

$$\sqrt{c^2 - v^2} \, ,$$

i.e. slower than the speed of light **c**.

The resultant speed over distance A_1B_1 is defined further in Chapter 7.

Triangle of Velocities as seen from the AFR

'v' represents the velocity of centres of photon bonding at Ⓐ Ⓑ and Ⓒ moving right in the AFR through positions A_1, B_1 and C_1.

'c' represents the speed in the AFR over the mean path of photon bonding along the hypotenuse.

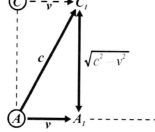

$\sqrt{c^2 - v^2}$ represents the resultant 2-way bonding velocity of photons between A_1 and C_1 calculated from the AFR triangle of velocities AC_1A_1.

Now, the mean resultant velocity of photons between A_1 and C_1, $\sqrt{c^2 - v^2}$, may also be written with respect to c, as $\sqrt{c^2 - v^2} \times c/c$, or $\sqrt{1 - v^2/c^2} \times c$.

Since $\sqrt{1 - v^2/c^2}$ will be a frequently occurring factor that decreases as velocity increases, let it be termed ϕ. Photons whose speed in the AFR is c, would then be seen by instruments stationary in the AFR to travel at a slower speed [15] over the perpendicular distance A_1C_1 at a resultant bonding velocity of

$$\phi c$$

Clock Rate Change

Let us now consider how this variation in the speed of light affects the rate, i.e. the measurement of time, of a moving clock. If such a clock relies on the sending and receiving of electromagnetic signals through the medium of the AFR, then we propose that when the clock moves through the AFR the clock will run slower. This seems to be the result from SR, but it is not for the same reason. Our proposal is that the *rate* of clocks and all processes[16] decrease *absolutely* due to motion. In other words, it is not time per se that changes as Einstein suggests in his space-time, but simply *the measuring instruments of time*, i.e. clocks record at a different rate when moving, and this has been tested experimentally.

We can illustrate this clock rate change by beginning with a simple example of measuring time using a *lightclock* that is moving through the AFR. A duration of time is measured by a 2-way flash of light to and from a reflector. From a calculation in the AFR, similar to the one above, the light travels at a slower velocity of ϕc, over a 2-way journey perpendicular to the motion. If the absolute distance travelled is s, the effective duration t, for a moving lightclock is longer at

$$t = 2s/\phi c.$$

Should the same lightclock have been at rest in the AFR let the absolute time of the journey be termed

$$t_0 = 2s/c.$$

Then:

$$t / t_0 = \frac{2s}{\phi c} \times \frac{c}{2s} \quad \text{or} \quad t = t_0 / \phi.$$

This is the familiar result of Time Dilation put in *SR terms* of a moving body where

$$t' = t / \phi.$$

However, for *framing measurements in absolute terms*, let t_0 be the duration of a timed event measured by a clock at rest, by which we mean stationary in the AFR. Then the duration of a timed event of a moving body measured by the same clock is t. Now, from the expression $t = t_0 / \phi$, above, we can say that the duration of an event t, occurring in the moving body recorded by the clock at rest in the AFR will take longer when compared with the absolute duration t_0, such that

$$t_0 = \phi t$$

This is the formula proposed for Clock Rate Change and is demonstrated by experiment where moving clocks and processes are seen to run slower by the factor ϕ, than when they were stationary.

However, this is not the *slowing* of time. Nor is it time dilation in relativistic terms, i.e. that time itself is something that changes. *Time itself is not changed by motion.* It is simply a consequence of the rate of processes and measuring instruments of time $t_0 = \phi t$, (of actual physical clocks) that are changed as a result of moving through the AFR. This causes clocks to record different information.

This leaves the way open not only for *absolute* definitions of time but also for a velocity of light that always *appears* constant since it is measured by clocks that change with motion.[17] Thus an application of this new result shows that the so called Einstein's Clock or Twin paradox vanishes.[18]

Einstein conceded in his Special Theory that a radical rethink of the nature of time was crucial, and indeed SR predicts the slowing of time. However, the comprehensive solution provided in this book is the non-relativistic classic conclusion with clocks that slow and lengths that decrease.

Length Decrease

We now propose that the effect of a body moving through a material AFR is that its length will decrease in the direction of travel. A complete description of quantisation and the uncertainty principle of a body at the sub-atomic level is highly complex because it involves the mutual interactions of many particles and demands more attention than the elementary argument that we will provide here.

We therefore take a simple example for illustrative purposes. Suppose that the stability of an ordinary sample of matter depends on chemical bonding explained in terms of contributions from constituent molecules. Let us then consider hypothetical centres in a solid body held in place by virtual photons of quantum molecular bonding that propagate between component particles. Neither Einstein, nor Lorentz of course, had the advantage of such knowledge.

We therefore propose to adopt a series of these hypothetical centres to represent points where bonding takes place by using a vector analysis in order to describe electromagnetic interactions of molecular bonding in a body.

Now, such a state would depend on the precise *duration* of bonding exchanges. These consist of multiple virtual photons communicating at the speed of light, so that the *average time taken* for quantum bonding photons to travel between centres is an important criterion in maintaining molecular stability. The arguments that require an analysis of the electrostatic forces within a body in motion then become unnecessary.

We must carefully note in the following discussion the importance of whether measurements are made by instruments that are stationary or moving in the AFR.

Consistency in reciprocity is equally important where instruments of a moving body could be considered as stationary and the AFR as moving.

Therefore let us select three hypothetical centres at points A, B, and C, in the figure below. They would represent points in a body that are perpendicular and parallel respectively to the direction of motion, between which bonding described above is resolved into directions AC and AB.

For a body at rest suppose distance AC is situated perpendicular to AB, and where AC and AB are absolute distances d_0 and l_0.

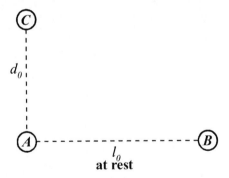

In motion, suppose distance A_1C_1 is termed d, and distance A_1B_1 is termed l.

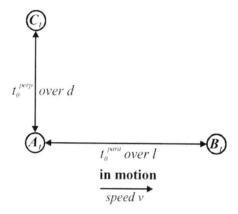

Let t_0^{perp} represent the resultant duration of a 2-way perpendicular bonding between A_1 and C_1.

Let t_0^{para} represent the resultant for a 2-way parallel bonding between A_1 and B_1.

Let t, be the duration of an event in a body in motion and t, be multiplied by the factor ϕ, for Clock Rate Change as $t_0 = \phi t$.

Each of these measurements is made by instruments at rest in the AFR which will view the figure above as a body moving to the right at speed v.

Now, photon bonding requires that the duration t_0^{perp}, is determined by the slow transfer of bonding at ϕc, over the 2-way perpendicular distance d.

We then see that the duration $t_0 = \phi t$, of 2-way journeys of bonding photons as multiplied by ϕ, for Clock Rate Change takes place in such a way that the status quo in the *perpendicular* direction is conserved where

$$t_0^{perp} = \phi \left(\frac{d}{\phi c} + \frac{d}{\phi c} \right) = \frac{2d}{c} \, .$$

The same factor for Clock Rate Change ϕ, is also applied in the case of 2-way *parallel* journeys of photons which communicate at velocities $c+v$, and $c-v$, over the length l, so that the term $t_0 = \phi t$, becomes

$$t_0^{para} = \phi \left(\frac{l}{c-v} + \frac{l}{c+v} \right) = \phi \left(\frac{2lc}{c^2 - v^2} \right).$$

Since $\phi^2 = \dfrac{c^2 - v^2}{c^2}$ we can write $\phi \dfrac{2l}{\phi^2 c}$ or

$$t_0^{para} = \frac{2l}{\phi c} \, .$$

Even though the factor for Clock Rate Change ϕ, is now applied to both equations, virtual photons travel in the direction parallel to motion at a *decreased overall speed* of ϕc. This must mean either that the absolute time t_0^{para}, increases or that length $2l$, decreases.

Interestingly, there is no trace of experimental evidence in a moving case to suggest that the *absolute time* of light travelling over parallel distances differs from perpendicular distances. *Thus, it must be that lengths decrease*.

Combining these two equations as a ratio demonstrates this conclusively, where

$$\frac{t_0^{perp}}{t_0^{para}} = \frac{2d}{c} \times \frac{\phi c}{2l} = \phi d / l.$$

It is here that we notice a similar ratio of times for the 2-way journey of light between a source and reflector that occurs in such experiments that Michelson and Morley conducted to try to detect an aether. From these experiments, there is consistent support for the fact that this ratio is a constant. At the advent of SR this seemed to indicate that there was either no aether, no motion, or that motion in the aether could not be detected and should be ignored.

However, we propose that the instruments of moving bodies as such are merely unaware of this ratio. This is because we have arrived at this ratio from a perspective, not only from an AFR, but also on the assumption that quantum photons in a rigid body that is moving in an AFR can travel between molecules at different speeds.

On this basis, we now show that the constancy of this ratio is interpreted in a radically different way from SR.

If $t_0^{perp} / t_0^{para} = K$ (a constant), then lengths l of bodies in motion will decrease as

$$l = \phi d / K.$$

Take a particular case *at rest* in the AFR. This will give $\phi = 1$, $d = d_0$, $l = l_0$, and $K = d_0/l_0$. For bodies in motion we can therefore also write

$$\phi d/l = d_0/l_0 \ .$$

Since changes in distances in the parallel direction only seem to be involved where a body is in motion, the expression $d = d_0$ also applies. By substituting d for d_0 in the equation above, lengths l, will be found to decrease as

$$l = \phi l_0.$$

This result now seems to conform to SR, but is in effect entirely different. The implication here is that when a body moves in the AFR, the physical decrease of actual lengths masks the motion of the body. Therefore, we can conclude that since velocities of perpendicular and parallel journeys of bonding photons vary by the factor ϕ, parallel lengths of physical objects will also vary by the same factor with respect to each other. This ensures that the ratio of times of the journeys remain constant as all experiments show.

Any length moving through the AFR undergoes a contraction not normally observed because the *measuring instruments contract correspondingly.* Thus, the distance appears unchanged. In the MM experiment this would compensate exactly for the expected difference in the velocity of light in the two right angled directions. The ratio of the velocities would thus appear to be unity.

This proposal, of course, is open to further examination. However, it does demonstrate a basis on which we can return to absolute measurements at rest in the universe, and offer explanations for problems accompanying SR, such as we will now see in a study of Simultaneity.

6

Simultaneity

Let us therefore examine Einstein's definition of Simultaneity more closely based on our discussion in terms of time.

Einstein stipulates as shown above that

> *"The two clocks are synchronous by definition if*
>
> $$t_B - t_A = t'_A - t_B \text{ "}$$

and that

> *"Based on experience, we further stipulate that the quantity*
>
> $$\frac{2\overline{AB}}{t'_A - t_A} = c$$
>
> *be a universal constant (the velocity of light in empty space)."*

If we simplify this second equation where $t_A = 0$, and let distance $2AB = 2l_0$, Einstein's sum of times for the 2-way passage of light between A and B, t'_A, can be written as

$$t'_A = 2l_0 / c.$$

Einstein implied from this that the times for the outward and return journeys must be the same, which disallows an AFR unless the system is *"a rest system"*, thus denying the existence of an aether. That said, Einstein's definition of Simultaneity remains ambiguous as practical knowledge of the value of B's clock t_B, is denied, and his theoretical conclusion is not free of contradictions since it is maintained without means of proof. However, let us now consider this equation viewed from the absolute position of the AFR.

Suppose the system of AB is *not* seen to be *"a rest system"* in Einstein's terms, but a system travelling 'east' at speed v, with respect to the AFR in which light travels at speed c. How do we synchronise A and B time, t_A and t_B, from an absolute position?

Let us use 2-way light signals in a moving system monitored by instruments at rest in the AFR.

Although the duration of the sum of times under *lengths decrease* above for the 2-way journey parallel to v was $t_0^{para} = 2l/\phi c$, can we now match this expression with Einstein's time embodying his absolute term, t'_A?

Since length AB decreases with motion, as $l = \phi l_0$, above, we can substitute the term l, for, ϕl_0 so that the expression $t_0^{para} = 2l/\phi c$, above, becomes $t_0^{para} = 2\phi l_0 / \phi c$. When simplified we then have

$$t_0^{para} = 2l_0 / c.$$

Next, we find above that since t'_A also equals $2l_0 / c$, we can forego the term $2l_0 / c$, and embody t'_A instead, so that

$$t_0^{para} = t'_A.$$

We now see that the overall time t_0^{para}, of our 2-way passage of light in a moving object is still the same that Einstein stipulated, *even though it does not originate from his "rest system".*

Furthermore, the speed of light for the two 1-way journeys of light, A to B and B to A, and their times, i.e. $2\phi l_0 /(c + v)$ and $2\phi l_0 /(c - v)$, are different, which Einstein completely disallowed.

Extraordinarily, Einstein's assumption masks the fact that the 1-way time and speed of light for AB *can be different.*

Note. We can only ever observe the 2-way overall average results of the passage of light, and that is the root of the problem for Relativity Theory.

7

The Speed of Light

As has been demonstrated by experiments, let us now show why the speed of light is always found to be the same *numerical* value c, masking the fact that its speed can vary for 1-way travel in moving objects.

In this discussion, we must be careful to note that instruments at rest in the AFR record measurements that are absolute, and that instruments moving with the body AB record measurements that are not absolute.[17]

Suppose AB is moving at speed **v** while light is transmitted from A to B and reflected back to A. The light travels parallel to **v** at speed **c** *in the AFR*. Due to motion it is received at B at speed **c** − **v** and at A at speed **c** + **v**.

$$A \underline{\qquad \phi l_0 \qquad} B$$
$$\longrightarrow v$$

First, let instruments that are stationary in the AFR determine the speed of light for AB. Distance AB is measured by instruments stationary in the AFR as contracted length $l = \phi l_0$. The absolute duration t_0, for the 2-way parallel beam of light reflected over AB will then be timed by a clock stationary in the AFR.

The total time t_0, for the 2-way parallel beam of light that travels from A to B and back at different speeds over shortened distance AB will then be equal to

$$t_0 = \frac{\phi l_0}{c-v} + \frac{\phi l_0}{c+v}.$$

Adding the two times gives

$$\frac{2\phi c l_0}{c^2 - v^2}.$$

Since $\dfrac{c^2 - v^2}{c^2} = \phi^2$, $c^2 - v^2$, can be replaced by $\phi^2 c^2$. Then the expression becomes

$$\frac{2\phi l_0}{c} \times \frac{c^2}{c^2 - v^2},$$

or

$$\frac{2\phi l_0}{\phi^2 c}.$$

Simplifying again gives us the time

$$t_0 = \frac{2l_0}{\phi c}.$$

However, since AB is moving, this expression of time for AB, which is now

$$t_0 = 2l_0 / \phi c ,$$

should be multiplied by ϕ, for clock rate change, so that it becomes $\phi \times (2l_0 / \phi c)$, or

$$2l_0 / c .$$

This gives the value of the time seen from the AFR but measured by a clock <u>moving with AB as</u>

$$t^{AB} = 2l_0 / c .$$

Next, instruments travelling with AB will be unaware of any movement through the AFR, and will measure the contracted length as l, which the AFR sees as being ϕl_0. However, rulers moving with AB contract in the same ratio as lengths being measured. Then instruments moving with AB but viewed from the AFR will obtain the *value* for the length AB as

$$\phi l_0{}^{AB} = \phi l_0 ,$$

and for the distance for the 2-way journey of light A to B and back as

$$2l_0{}^{AB} = 2l_0 .$$

So, according to the units of length and time on the measuring instruments that are moving, the value of the 'average' speed of light for the 2-way travel (total distance over total time) is $2l_0 \div 2l_0 / c = c$, or

$$2l_0{}^{AB} / t^{AB} = c.$$

The constant *numerically the same* value of c is derived here by instruments moving with AB using units of length and time that are affected by length decrease and clock rate change. Thus, although the speeds ($c-v$ and $c+v$) of 1-way journeys of light between A and B are different, the <u>overall</u> *average* speed is constant for this and any double journey. Without considering any reciprocal view of the AFR, experiments will always show this result.

We stress again this important observation: the speed of light averaged over a 2-way journey to a reflector and back is found to have the same *value* that we denote by c, whatever the speed of the moving instruments. This is because the measurements of time and length change with motion, when compared with absolute measurements made from instruments stationary in the AFR. The duration of time is thus longer as t^{AB}, and lengths are shorter at $l_0{}^{AB}$, but this is not perceived by the instruments in motion.

Interestingly this calculation produces the same *value* for the speed of light.

8

Addendum

Reciprocity in the AFR

Let us study reciprocity of time from the viewpoint of S in the diagram. S is moving east at v, with respect to S_0, which is stationary in the AFR.

Next, let a light beam of S_0 travel perpendicular to v, over the 2-way distance $2d$ [ABBA] in time $t_0=2d/c$.

While S is moving east, S sends a 2-way beam of light perpendicular to v, over the same distance $2d$ [C'B'B'A'] and this takes the period $t=2d/c$. According to the moving clock of S the measured value of t, is the same as t_0 but with instruments that are calibrated differently due to motion.

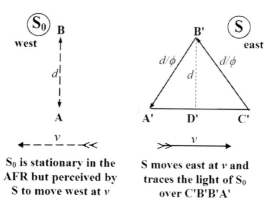

S₀ is stationary in the AFR but perceived by S to move west at v

S moves east at v and traces the light of S₀ over C'B'B'A'

61

As S travels east, S sees S_0 appear to travel west at v. S then traces the path of the light beam of S_0 [ABBA] and, measured on instruments carried by S, sees the light apparently move over the route [C'B'B'A']. Using Pythagorean Theory, S finds that this beam travels at velocity c over distance $2d/\phi$. The clock of S then calculates that the light beam of S_0 travels over distance $(2d/\phi)$ [C'B'B'A'] in the time $2d/\phi c$. [Note that ϕ decreases as v increases].

As S passes the clock face of S_0, S sees that the light beam of S_0 travels at the slow speed of ϕc, over the distance $2d$, in the slow time $2d/\phi c$. Then, using the term t^R, S records the time for the S_0 event and registers it as reciprocal time running slow at

$$t^R = 2d/\phi c.$$

<u>S declares by classical means that the clock of</u>
<u>S_0 stationary in the AFR is reciprocally slow.</u>

Remark: This result does not mean that S_0 reads his clocktime or the velocity of his light as slow. It simply signifies that while S_0 remains stationary in the AFR, and while S passes S_0, S *perceives* that S_0 clocktime is reciprocally slower than that of S clocktime.

Using this result, there are similar lines of thought that could be followed in the AFR to show that reciprocity of mass increase and length decrease is also possible.

A classical derivation of $E=mc^2$

It is appropriate to draw brief reference to one of the most well known equations usually associated with the Theory of Relativity.

We do this, not because it is essential to the main purpose of this book, but because the equation that links energy to mass is thought to come from the Theory of Relativity. Being so widely accepted, it seems to verify the theory and be in opposition to any challenge, such as that which we are bringing in this book.

It may surprise the reader to learn that the equation $E=mc^2$ is built on experimental evidence that was later taken up by Einstein and brought into his developing theories. Indeed, the mathematics behind the equation is much simpler than most people realise, once the experimental evidence is accepted.

We therefore propose that a simple classical argument shows that Einstein should not have claimed that this famous equation stemmed from Special Relativity. His understanding of the implications brought profound consequences but the equation itself does not depend on Special Relativity.

The classical derivation below is actually built on a short introduction to Einstein's relativistic explanation and can be found in:

Physics for the Enquiring Mind by Eric M Rogers. [19]

Derivation of $E=mc^2$

This short derivation, due to Einstein, uses the experimental knowledge that when radiation with energy E joules is absorbed by matter, it delivers a momentum of E/c *kg.m/sec.*

(Experiment shows that the PRESSURE of radiation on an absorbing wall is ENERGY-PER-UNIT-VOLUME of radiation-beam. Suppose a beam of area A, falls on an absorbing surface head-on. In time Δt, a length of beam $c.\Delta t$ arrives.

Then MOMENTUM delivered in Δt

$$= force \times \Delta t$$

$$= pressure \times A \times \Delta t$$

$$= (energy/volume) \times A \times \Delta t$$

$$= (energy/A \times c \times \Delta t) \times A \times \Delta t$$

$$= energy/c$$

This also follows from Maxwell's equation)

A. Einstein.

To continue therefore, it follows from above that the *momentum* increase, per second, or *mv*, where v is the increase speed becomes

$$mv = energy/c.$$

That said, Einstein further performed a simple relativistic "thought experiment" based on this short introduction to derive the equation $E=mc^2$. In a similar manner, we also propose a simple but classical equivalent.

We know from the above, that experimental evidence shows that when radiation with energy E joules is absorbed by matter, it delivers momentum ***mv=energy/c* or**

$$E/c \ kg.m/sec.$$

Suppose a block of matter at rest in the AFR absorbs $\frac{1}{2}E$ from due East and $\frac{1}{2}E$ from due West. Then we make measurements with instruments belonging to the block where the energy is totally absorbed by the block.

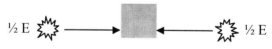

block with its instruments

Suppose that the mass equivalent of the energy E is m. Since the speed of its arrival is c, the momentum absorbed is mc *per sec*, which gives $mc=E/c$. Rearranging, the mass increase of the block due to energy or energy due to mass increase becomes

$$E=mc^2$$

Extensions of this simple thought experiment to moving blocks will require a little more maths but will achieve the same result. Thus, we see from this clear illustration that Relativity Theory *is not necessary* for the knowledge of the mass equivalence of energy. It relies on experimental evidence already known and Einstein, at the time, merely demonstrated that he could incorporate this equation into his developing theory of relativity.

Addition of Velocities

When considering Addition of Velocities this is assessed in Newtonian terms. If a disturbance is made in the AFR, radiation travels in opposite directions at the combined speed $2c$. Additions and subtractions of velocities in other situations follow likewise.

We would refer again briefly here to the subtle elegance of the Lorentz Transformations, and suggest that this *elegance* is beguiling and is *a root of the problems in SR*.

<u>We submit, therefore, that Lorentz's equations should be rejected.</u>

9

Conclusions

We have proposed that physics return to the consideration of an Absolute Frame of Reference in absolute space, and have shown how SR is incompatible with such a model. Einstein's suppositions were inexact when he assumed that an AFR could be ignored. It is intriguing that this fact has been obscured for over a century. This is because the 2-way 'average' speed of light travelling between a source and reflector merely agrees with the suppositions of his theory. This is advanced while hiding the inequality of the 1-way journeys of the light in a body moving in an AFR.

In arguing these points of making a variable speed of light a distinct possibility, we have raised important new perspectives beyond standard theory which seem close to relativistic concepts, but are so different as to imply significant redefinition. By this, we imply the existence of a form of aether that not only makes a variable speed of light a possibility but also opens the way to further studies beyond standard theory.

While the way is also open of course for other such proposals for models of the universe independent of the need to incorporate Special Relativity, we believe that, by using an Absolute Frame of Reference, this book presents a logical break-through beyond it.

That said, a new book is planned, entitled *Aether Physics—a new approach to the nature of the universe*. This investigates how the universe may be understood more clearly from the physics, space and time from which it is deemed to have been made using the existence of an Absolute Frame of Reference.

Table of issues affected by increase in velocity

Issue	Relativity Theory	Our Perceptions
Speed of Light	Einstein assumes this is constant in any direction.	We say constancy is only in *averaged* two way travel.
Time	Einstein says *time* changes.	We say *clocks in motion* record time differently due to their interaction with the AFR.
Length	Einstein says *length* changes.	We say *actual physical material and rulers* change .
Simultaneity	Einstein assumes the time for the outward and return journey of light *is the same.*	We say the time for the outward and return journeys of light *can be different* and this permits an AFR.
Mass	Einstein says *mass* increases, and that the equation $E=mc^2$ stems from Relativity.	We say mass increases but not for the same reason, and that $E=mc^2$ has a classical origin.
Absolutes	Einstein does not refute.	We say *there are absolutes* of mass, time and space.
Medium of the AFR	Einstein *does not deny* its existence.	We say *it does exist* but may not be detectable for the reasons stated.

Experiments

Our challenge to the foundational assumptions of SR opens the way to several reappraisals of what we understand of the physical universe. It makes way for a range of experiments, explained in our proposed new book *Aether Physics*, on inertia, gravitation, electromagnetism and matter based on the assumption of an AFR.

In short, we propose that among such experiments these could possibly:

- reveal a change in the velocity of light passing through a discharge or Faraday Cage thereby confirming a different interpretation for the red shift of distant starlight;

- offer an alternative explanation for the CMB (Cosmic Microwave Background) by reassessing the nature of its wavelengths;

- demystify the incompatibility of relativity theory with the quantum phenomenon through using a fine tuned double slit experiment;

Endnotes

[1] Stachel, John, *Einstein's Miraculous Year*, edited and recently revised, Princeton University Press, 1998, pp.123-160.

[2] Eric M Rogers *Physics for the Enquiring Mind*, p491, Princeton University Press, 196. Einstein's basic principle of being realistic says that, *"Where the answer is impossible, the question is a foolish one"*.

[3] Einstein A. *Ideas and Opinions*, Crown Publishers, New York, 1954. *"Morality as we understand it is not a fixed rigid system. It is rather a point of view from which all the questions that crop up in life can and should be examined. It is an endless task; a permanent feature guiding our judgement and inspiring our behaviour"*. The widespread view now admits that, *"Everything is relative"*.

[4] A useful summary of this experiment is found in French, A.P., *Special Relativity* , Massachusetts Institute of Technology 1968, pp.51-56.

[5] These problems became such classic difficulties that scientists began to ignore or deny the existence of an aether. Whether or not it existed ceased to be an issue when the principles of relativity were conceived and attention was diverted from further investigations to detect it. It nevertheless seems it can still resolve problems that relativity has brought. These are those that occur in the meaning and measurement of time, its difficulties with quantum theory, electrodynamics,

singularities, the red shift of light and the origin and extent of the universe.

[6] *Einstein's Miraculous Year*, p124.

[7] *Einstein's Miraculous Year*, p93.

[8] *Einstein's Miraculous Year*, p124.

[9] Absolute rest is the state in space from which the velocities of all bodies in the universe are assessed whether detectable or not.

[10] Refer to French A.P. *Special Relativity* , p80.

[11] *Einstein's Miraculous Year*, p126.

[12] For instance, Peter Bergmann, *McGraw-Hill Multimedia Encyclopaedia of Science and Technology*, section 'Space-time' gives a useful description.

> "In accordance with the Lorentz transformations, both the time interval and the spatial distance between two events are relative quantities, depending on the state of motion of the observer who carries out the measurements. However, a new absolute quantity takes the place of the two former quantities. It is known as the invariant, or proper, space-time interval *t*, and is defined by the equation, where τ is the ordinary time interval such that,

$$\tau^2 = T - \frac{1}{c^2} R^2,$$

and R, is the distance between the two events, and c, the speed of light in empty space. Whereas T and R are different for different observers, t, has the same value. Conventionally, in terms of absolute definitions, time is measured in days, which can be subdivided into hours, minutes and seconds. The absolute measure of time until the 1960s is that a day is measured by one rotation of the earth as observed through the position of the sun. On the earth, clocks are calibrated to this. If a clock moves and its rate of recording changes, it has ceased to measure time correctly. If however, someone defines time according to its rate of recording, then time becomes a phenomenon of relativity and so is redefined."

[13] *Einstein's Miraculous Year*, p127.

[14] Here we note that "Time Dilation" is a relativistic term of SR. Later we shall return to the argument that this is misleading because time per se does not change. It is the *Clock Rate* of the measuring instruments that changes, not time. For this reason and because we propose a return to absolute measurements we will use the term "Clock Rate Change" instead of "Time Dilation", together with time rate factor, ϕ, as in equation $t=t_o/\phi$. We propose that this affects all processes. For similar reasons we will use the term "Length Decrease", as in equation $l=\phi l_o$. Though this may seem pedantic, we are concerned to avoid the use of relativistic terms.

[15] To help understand this important point – a slow velocity of light – we can illustrate this concept with a simple example. Suppose in our diagram a child's ball rolls at speed V, from side to side over AC, and back in the cabin of a ship at anchor. While steaming through the water at velocity v, the ball has a speed of its own over A_1C_1 and back of

$$2 \times \sqrt{V^2 - v^2} \; .$$

However, to instruments on a stationary barge nearby, the ball traces an apparent zigzag over A_1C_1 and back. Since the completed journeys take the same time as observed from the two perspectives, but over different distances, this must result in different speed measurements. This is relative motion in the normal Newtonian sense. Now, suppose we consider light. The only difference is that photons travel in the medium at a fixed maximum speed of c. Thus, since the two photon-journeys will likewise take the same time, the speed over the perpendicular distance A_1C_1 will appear from the perspective of the AFR to be slower at ϕc. Neither the child nor the solid body of course, will notice any difference in their own speeds, as we show in Chapter 7.

[16] We recognise that there are various kinds of timing mechanisms, e.g. atomic clocks and pendulum clocks. While agreeing that the slowing of clocks is a physical fact that has been tested, it remains to be seen if all timing mechanisms and processes run slow, at the same rate. Interestingly, if they do not, this would enable us to detect motion through the AFR. However, here we are simply taking the example of a lightclock for illustrative

purposes. In any event, one form of timing device, e.g. an atomic clock or lightclock would be used as the datum for calibrating all clocks in a moving system. Any difference in slowing for a variety of types of clocks would thereby be masked.

[17] Based on the assumption that such measurements can be accurately made absolute measurements i.e. t_o, and l_o, are those apparent to instruments presumed to be at the absolute rest position in the AFR.

While we realise that such concepts are hypothetical, our argument is nonetheless based on this hypothesis. Thus, the two photon-journeys likewise take the same time over the *perpendicular* distance A_1C_1 and will appear from the perspective of the AFR to be slower at ϕc. They are thus unaware of their motion with respect to the AFR and do not agree with instruments already at rest in the AFR. Resulting measurements by instruments at rest in the AFR are those measurements based on the reaction of a body assumed to be moving in the AFR, i.e. $t=t_o/\phi$ and $l=\phi l_o$. We need to distinguish these carefully, since moving instruments change when in motion.

[18] As an example we quote from Wikipedia, the free encyclopedia en.wikipedia.org/wiki/Twin paradox. (Accessed 12 August 2008.)

> 'In physics, the twin paradox is a thought experiment in Special Relativity in which a person who makes a journey into space in a high speed rocket will return home to find he or she has aged less than an identical

twin who stayed on Earth. This result appears puzzling, since the situation seems symmetrical, as the latter twin can be considered to have done the travelling with respect to the former. Hence, it is called a "paradox" . In fact, there is no contradiction and the apparent paradox is explained within the framework of relativity theory, that only one twin has undergone acceleration and deceleration, thus differentiating the two cases. The effect has been verified experimentally using precise measurements of clocks flown in aeroplanes.'

The 2-fluid theory however, has no problems with this paradox, since the note with respect to Reciprocity shows that measurements taken from the viewpoint of being stationary in the AFR are not changeable but absolute. However, measurements when taken from the viewpoint of S, that is moving, vary.

[19] Eric M Rogers, *Physics for the Enquiring Mind*, p491, Princeton University Press, 1960.

Index

T

U

V